Reptile Adventure

Gila Monsters

MARK WEAKLAND

BLACK RABBIT BOOKS

Bolt is published by Black Rabbit Books
P.O. Box 227, Mankato, Minnesota, 56002
www.blackrabbitbooks.com

Alissa Thielges, editor; Michael Sellner, designer and photo researcher

Library of Congress Cataloging-in-Publication Data
Names: Weakland, Mark, author.
Title: Gila monsters / Mark Weakland.
Description: Mankato, Minnesota: Black Rabbit Books, [2024] | Series: Reptile adventure | Includes bibliographical references and index. | Audience: Ages 8-12 | Audience: Grades 4-6
Identifiers: LCCN 2023001221 (print) | LCCN 2023001222 (ebook) | ISBN 9781623106430 (library binding) | ISBN 9781623106492 (ebook)
Subjects: LCSH: Gila monster—Juvenile literature.
Classification: LCC QL666.L247 W43 2024 (print) | LCC QL666.L247 (ebook) | DDC 597.95/952--dc23/eng/20230412
LC record available at https://lccn.loc.gov/2023001221
LC ebook record available at https://lccn.loc.gov/2023001222

Printed in China

Image Credits

Alamy/BIOSPHOTO 16–17, Alamy Daniel Borzynski 19, Alamy Genevieve Vallee 1, Alamy Ger Bosma 3, 32, Alamy imageBROKER 21 (b); Alamy/Nature Picture Library cover, 5 (Gila), 12–13 (t), Rick & Nora Bowers 21 (t), Ron Niebrugge 4–5 (fox); Dreamstime/Heather Mcardle 10, Isselee 10–11; Flickr/Cameron Rognan 23 (t, inset), Flickr Lon&Queta 24; Getty/David Mosher/EyeEm 27 (br), Getty Peter Finch 23 (main), Getty Rick (RaVen) Hirschl 18; iStock/Juanmonino 27 (tl); Shutterstock/Cheri Alguire 26, Evelyn D. Harrison 20, Gwoeii 14–15, ianakauri 24–25 (icons), Milan Zygmunt 28–29, reptiles4all 8–9, 31, Sibons photography 6–7, William Silver 13 (b)

Contents

CHAPTER 1

A Bite in the Dark

A gray fox pads through the night. Sniffing, it comes upon a strange lizard. The lizard hisses a loud warning. But the fox steps closer. The lizard clamps its jaw onto the fox's foot. **Venom** flows from the lizard's jagged teeth. Is this the end for the fox?

Gila is pronounced HEE-la.

Shy Creatures

Gila monsters are lizards with deadly bites. People may fear them because of the venom. But Gilas are quiet creatures. They avoid humans and other large animals. It would be rare to see one in the desert.

PARTS OF A GILA MONSTER

FAT TAILS

How Big Is a Gila Monster?

Large Lizards

Gila monsters grow to be about 20 to 24 inches (51 to 61 centimeters) long. A fat one can weigh up to 4 or 5 pounds (1.8 to 2.3 kilograms). They're the largest lizards in the United States.

CHAPTER 2

Where They Live

Gila monsters only live in North America. Many are found in the southwestern United States. They also live in northwestern Mexico. Both areas are dry. Much of them is desert.

Gila monster's name comes from Arizona's Gila River. People first found them at this river.

WHERE GILA MONSTERS LIVE

ASIA
EUROPE
AFRICA
AUSTRALIA

Lazy Days

Gila monsters spend a lot of time quietly lying in dry **burrows**. Once in a while they leave to **bask** in the sun and feed. These lizards don't travel far. Most stay within a 1 square mile (2.6 sq km) area.

If a Gila bites a human, the bite is painful. It rarely results in death, though.

Cactus Climbers

These lizards walk high on their short legs. As they move, they hold their tails high off the ground. They chug along at a slow and steady pace. Gilas might not walk fast but they are good climbers. They might climb cacti, looking for bird nests.

CHAPTER 3

What They Eat

Gilas use their sense of smell to hunt. To track their **prey**, they flick their forked tongues out. The tongues pick up scents in the air. These slow lizards must sneak up on their meals.

Their prey includes baby birds and mice. They also eat insects and worms. Sometimes they snack on **carrion**.

Gilas don't need to use much energy. They also store fat in their tails. They don't need to search for food very often. Some eat only three or four times a year.

CHAPTER 4

Gila Monster Family

Gila monsters spend most of their time alone. But sometimes one shares its burrow with another. At other times, males search for females to **mate** with. When a female is ready to lay eggs, it digs a shallow hole. Then the lizard lays 3 to 13 eggs. It covers them with a little dirt and sand. The heat of the sun **incubates** them.

Males compete for mates by wrestling. The biggest and strongest monster wins.

Comparing Lengths

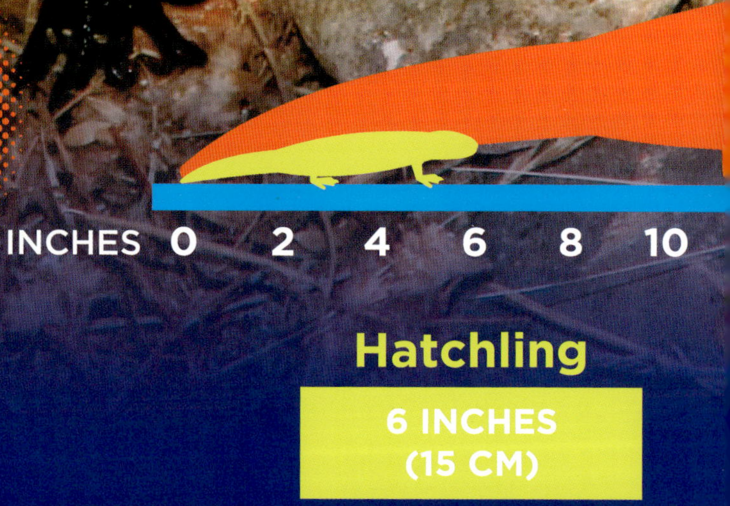

Growing Up

It takes about four months for babies to grow in their eggs. Then they're ready to hatch. The babies crawl out of the holes. Now they are on their own. Hatchlings are only a few inches long. They look like tiny adults.

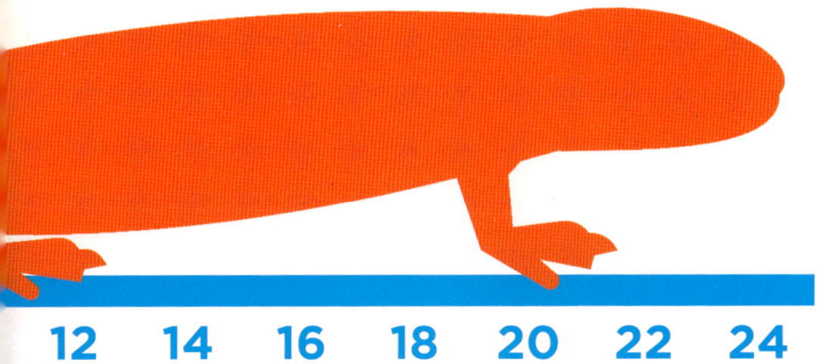

Adult

20 TO 24 INCHES (51 TO 61 CM)

Lizards Lives Threatened

Like many reptiles, these lizards are threatened. Birds like hawks eat them. So do foxes and coyotes. But their main **predators** are humans. Some people kill them out of fear. Some die when their habitat is cleared for fields and homes. Cars crush the lizards as they cross roads too. People need to look out for these **unique** creatures.

Gila Monster

Threats

humans

habitat destruction

• • • hawks

By the Numbers

about 20 years

life span in the wild

1 to 1.5 POUNDS
(0.5 to 0.7 kg)

how much food a Gila monster can eat at once

1 to 1.5 MILES
(1.6 to 2.4 kilometers)
PER HOUR

TOP SPEED

bask (BASK)—to lie or sit in the sun and enjoy it

burrow (BUR-oh)—a hole in the ground made by an animal for shelter or protection

carrion (KAR-ee-uhn)—the flesh of dead animals

incubate (in-kyuh-BEYT)—to keep eggs warm

mate (MAYT)—to join together to produce young

predator (PRED-uh-tuhr)—an animal that eats other animals

prey (PRAY)—an animal hunted and killed for food

unique (yoo-NEEK)—very special or unusual

venom (VEH-num)—a poison made by animals used to kill or injure

BOOKS

Maloney, Brenna. *Reptiles.* Fast and Slow Animals. New York: Children's Press, an imprint of Scholastic Inc., 2023.

Tobler, Elise. *Gila Monsters Have a Deadly Bite!* Reptiles Rock! New York: Enslow Publishing, 2021.

WEBSITES

Banded Gila Monster
stlzoo.org/animals/reptiles/lizards/banded-gila-monster

Gila Monster
sdzwildlifeexplorers.org/animals/gilamonster

Gila Monster
nationalzoo.si.edu/animals/gila-monster

INDEX